U.S. Military Forces

U.S. AIR FORCE

JULIA GARSTECKI

BLACK RABBIT BOOKS

Bolt is published by Black Rabbit Books
P.O. Box 3263, Mankato, Minnesota, 56002.
www.blackrabbitbooks.com

Jennifer Besel, editor; Michael Sellner, interior designer;
Catherine Cates, cover designer; Omay Ayres, photo researcher

Library of Congress Cataloging-in-Publication Data
Names: Garstecki, Julia, author.
Title: U.S. Air Force / by Julia Garstecki.
Description: Mankato : Black Rabbit Books, [2021]
Series: Bolt. U.S. military forces | Includes bibliographical references and index.
Audience: Ages 8-12 | Audience: Grades 4-6
Summary: "Learn about the powerful soldiers who make up the U.S. Air Force, including what they do and the training it takes to become one"— Provided by publisher.
Identifiers: LCCN 2019033063 (print) | LCCN 2019033064 (ebook)
ISBN 9781623102968 (hardcover) | ISBN 9781644663929 (paperback) |
ISBN 9781623103903 (ebook)
Subjects: LCSH: United States. Air Force—Juvenile literature. | Airmen—United States—Juvenile literature.
Classification: LCC UG633 .G295 2021 (print) | LCC UG633 (ebook)
DDC 358.400973—dc23
LC record available at https://lccn.loc.gov/2019033063
LC ebook record available at https://lccn.loc.gov/2019033064

Image Credits

af.mil: J.M. Eddins Jr. / U.S. Air Force, 20 (map); Master Sgt Jack Braden/U.S. Air Force, Cover (pilot); Master Sgt. John R. Nimmo Sr. / U.S. Air Force, 18–19; Master Sgt. Keyonna Fennell, 10 (t); Staff Sgt. Nicholas M. Byers / U.S. Air Force, 15 (b); Trevor Cokley / U.S. Air Force, 16–17 (main); U.S. Air Force, 1, 3, 6 (t & b), 8 (inset), 18, 20 (all sidebar images), 28–29, 32; afrc.af.mil: Lt. Col. Leslie Pratt / U.S. Air Force, 24–25; army.mil: Markus Rauchenberger, 12–13; commons.wikimedia.org / af.mil: Master Sgt. Robert Wieland, 8–9; defense.gov: Newsroom, 6 (c); iStock: freestylephoto, 31; mbda-systems.com: MBDA, 22–23; missiledefenseadvocacy.org: MDAA, 26; schriever.af.mil: Airman 1st Class Andrea A. S. Williamson / U.S. Air National Guard, 15 (t); Shutterstock: 19srb81, 20 (t); Ivan Cholakov, 4–5; ledokolua, 16–17 (bkgd); Neuevector, Cover (wings); Every effort has been made to contact copyright holders for material reproduced in this book. Any omissions will be rectified in subsequent printings if notice is given to the publisher.

Contents

CHAPTER 1

To Protect and Serve

A B-52 airplane soared above Afghanistan. U.S. Air Force pilots steered the plane over a **terrorist** training area. They released a bomb. The bomb destroyed the enemy's buildings. Then the pilots zoomed off toward another target.

BUILDING
airfields

stopping
CYBERATTACKS

studying
WEATHER

Important Roles

Bombing enemy areas is just one of the Air Force's jobs. The force also builds runways and airfields. Some soldiers stop **cyberattacks**. Others study the weather.

U.S. Air Force soldiers are **stationed** all around the world. They do many jobs. But they are always working to keep the United States safe.

In 2019, the U.S. Air Force had 327,315 members.

CHAPTER 2

Missions

The U.S. Air Force does many kinds of missions. Air strikes are one of the most common. Soldiers on the ground find enemy areas. They send the locations to pilots. Then pilots fly in and strike from above.

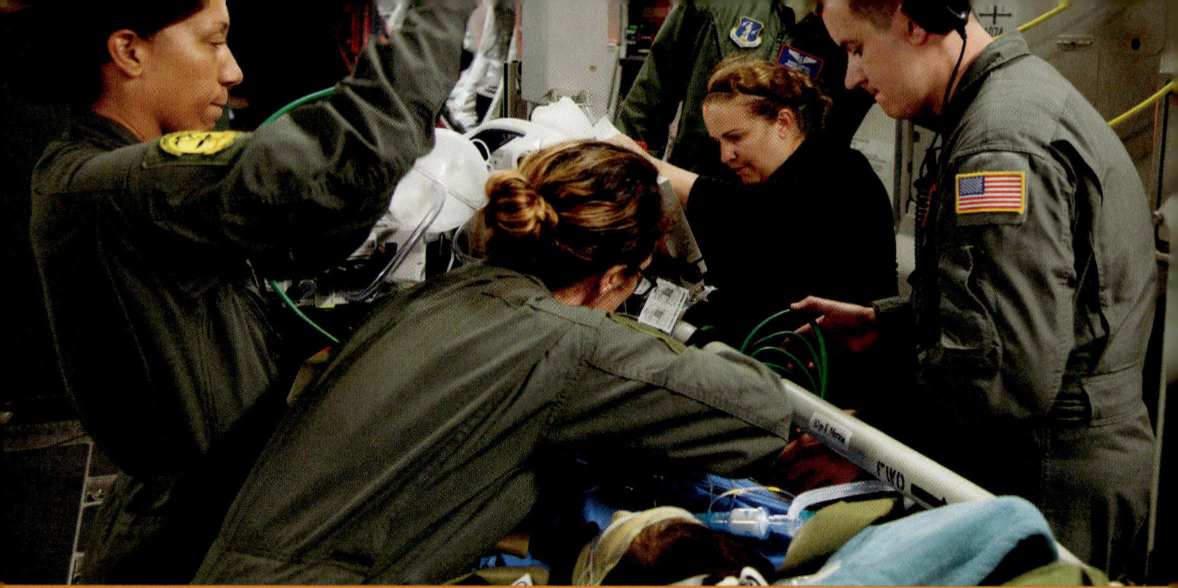

U.S. Air Force to Guatemala

To the Rescue

Airplanes are the fastest way to travel. With that speed, Air Force soldiers are also sent on rescue missions.

In June 2018, a volcano erupted in Guatemala. The U.S. Air Force was called to help. An Air Force medical team flew to the country. They rescued six children burned in the eruption. They flew the children and their families to the United States for treatment.

RAPPELLING

Aircraft can't always land safely on the ground. Some missions need airmen to rappel from helicopters.

LEATHER GLOVES
SAFETY BELT
ROPE

CHAPTER 3

Training and Jobs

New Air Force soldiers aren't sent on missions right away. It takes training to be ready. Every soldier must start with basic training. They learn to use weapons and work as teams. New soldiers are also tested physically.

PHYSICAL
TRAINING

WEAPONS
TRAINING

BASIC TRAINING

Basic training takes eight and a half weeks. Here are just some of the things soldiers do each week.

week 1
saluting directions and physical training

week 2
Air Force history, weapons handling, and physical training

week 3
physical training testing and cyber awareness

week 4
team building and anti-terrorism protection

introduction to Air Force combat

week 5

week 6

defense against different kinds of attacks

combat arms training and mission drills

week 7

week 8

get job assignment and graduation

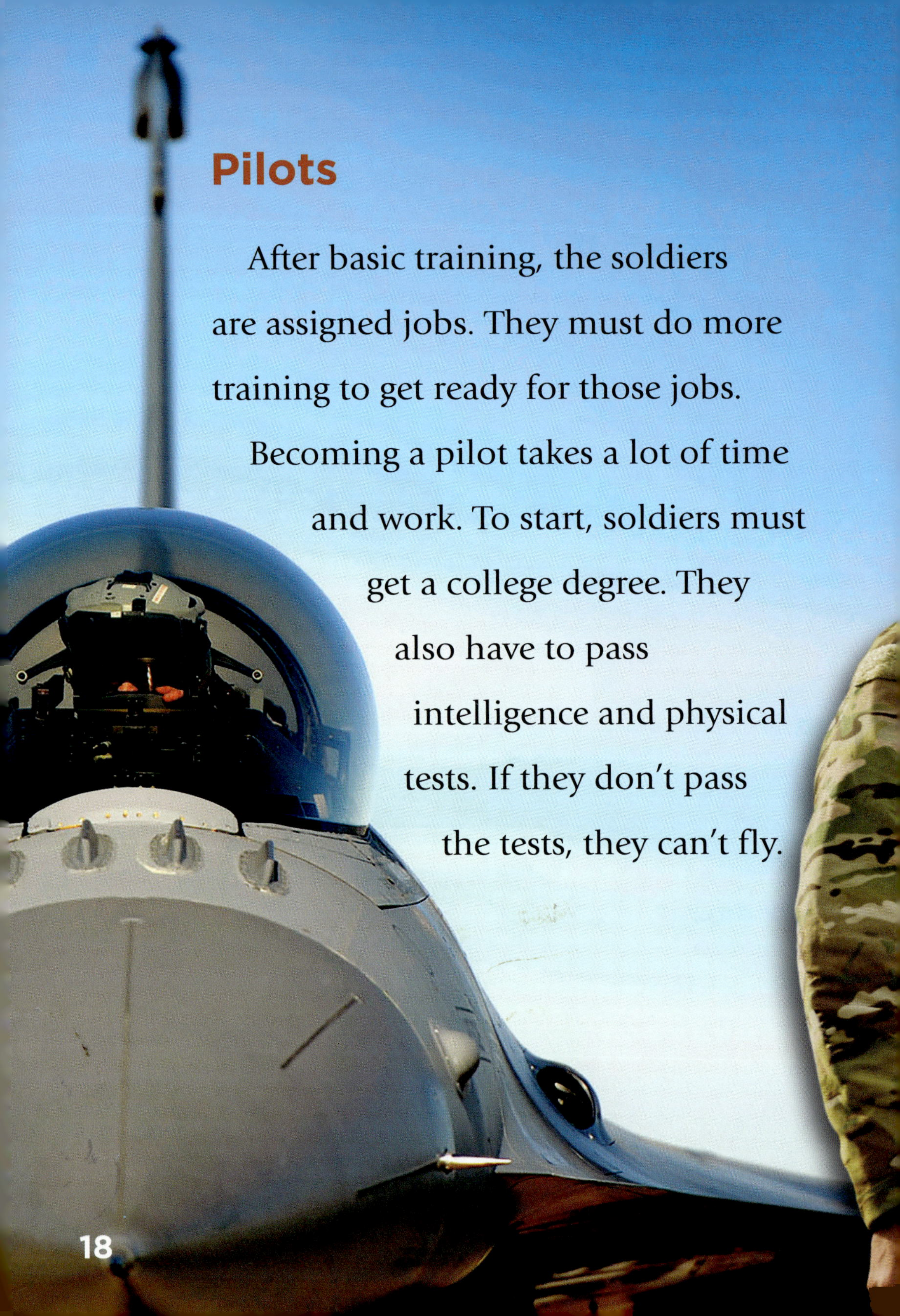

Pilots

After basic training, the soldiers are assigned jobs. They must do more training to get ready for those jobs.

Becoming a pilot takes a lot of time and work. To start, soldiers must get a college degree. They also have to pass intelligence and physical tests. If they don't pass the tests, they can't fly.

Watching the Weather

Imagine starting a mission when a storm strikes. That could be dangerous. That's why the Air Force has weather specialists. These soldiers learn to study weather patterns. They tell teams if storms are coming.

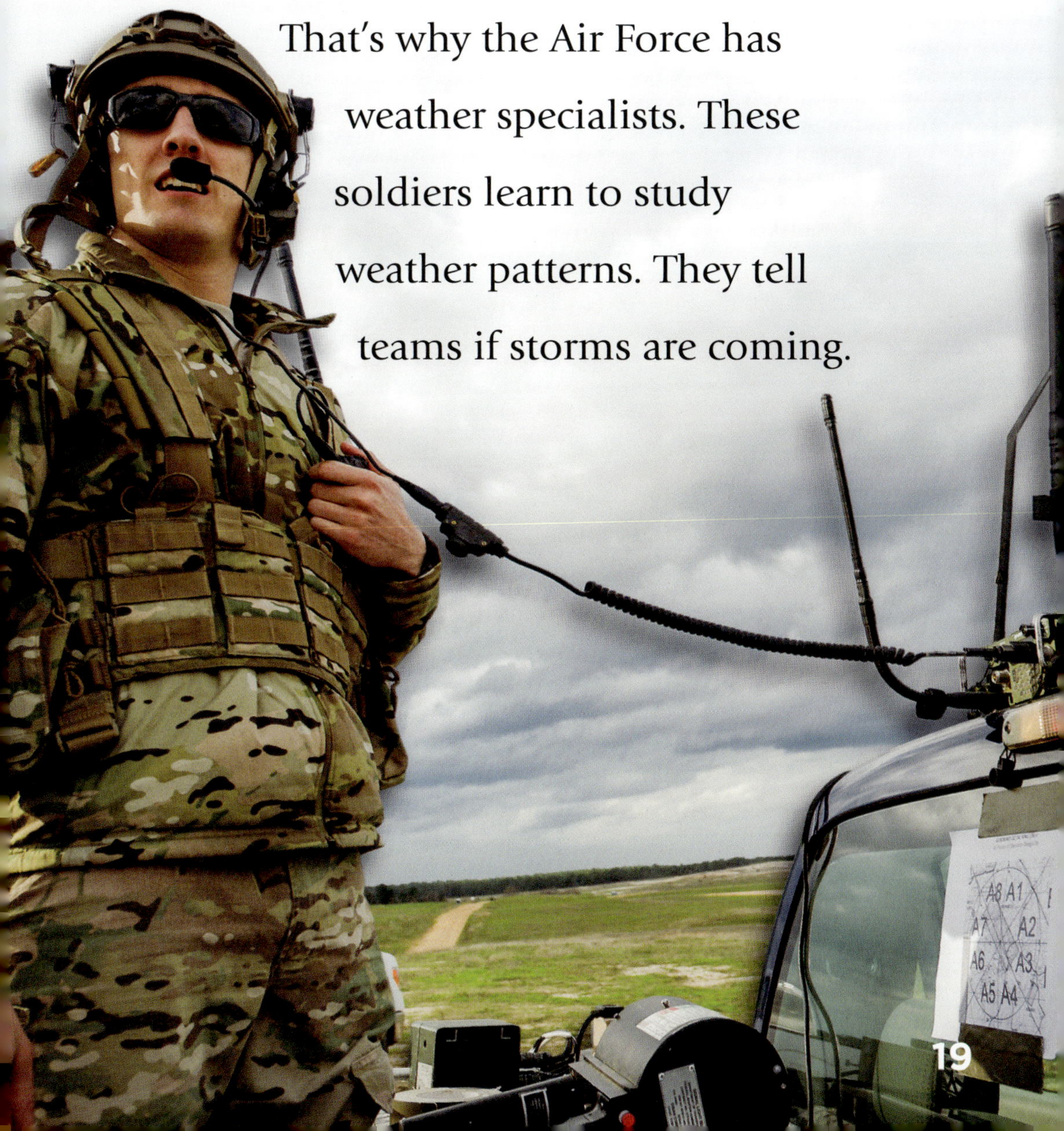

Other Jobs in the Air Force

weapons handling

airplane maintenance

emergency management

medical doctors and nurses

Tech Training

Cybersecurity is another important Air Force job. Weapons and planes use technology. **Hackers** could stop weapons from firing or send planes to the wrong places. Soldiers train to stop hackers from causing harm.

CHAPTER 4

Gear and Weapons

The U.S. Air Force uses high-tech tools to do its jobs. It's best known for the airplanes. The U.S. Air Force has some of the most advanced aircraft in the world. The force has more than 5,000 aircraft to use.

Total Counts of Some U.S. Aircraft

2,016
1,181
157
186
2,000
1,500
1,000
500
0
2019 total
bombers
fighters
training aircraft
helicopters

MQ-9 REAPER DRONE

By the Numbers

The Air Force uses **drones** too. The MQ-9 Reaper carries heavy weapons. It also records enemies with cameras.

NUMBER OF CAMERAS

up to **368**

TOP SPEED

376 miles

(605 kilometers) per hour

Lasers

The Air Force is working on new weapons too. Air Force researchers are developing high-powered lasers. Lasers could be used to destroy incoming missiles. They could also be fired at enemy aircraft.

Keeping the United States Safe

Air Force soldiers attack enemies. They stop hackers. And they research new weapons. No matter what they do, they have one goal. They are always on call to help the United States.

cyberattack (SY-bur-uh-tak)—an attempt to gain illegal access to a computer or computer system for the purpose of causing damage or harm

drone (DROHN)—an unmanned aircraft or ship guided by remote control or onboard computer

hacker (HAK-uhr)—a person who illegally gains access to and sometimes makes changes within a computer system

rappel (ruh-PEL)—to descend from something by sliding down a rope

station (STAY-shun)—to assign to or set in a post or position

terrorist (TER-ur-ist)—a person who uses violent acts to frighten people in order to achieve a goal

LEARN MORE

BOOKS

Boothroyd, Jennifer. *Inside the US Air Force.* US Armed Forces. Minneapolis: Lerner Publications, 2018.

Sherman, Jill. *U.S. Air Force.* Serving Our Country. Mankato, MN: Amicus, 2019.

Williams, Heather DiLorenzo, and Warren Rylands. *Air Force.* Eyediscover. New York: AV2 by Weigl, 2019.

WEBSITES

The Official Home Page of the U.S. Air Force
www.af.mil/

United States Air Force
www.youtube.com/user/AFBlueTube

U.S. Air Force
www.airforce.com

INDEX